KU-511-149

234239

Bevy of Beasts : Enchanting Animals
of Borneo, Belize and

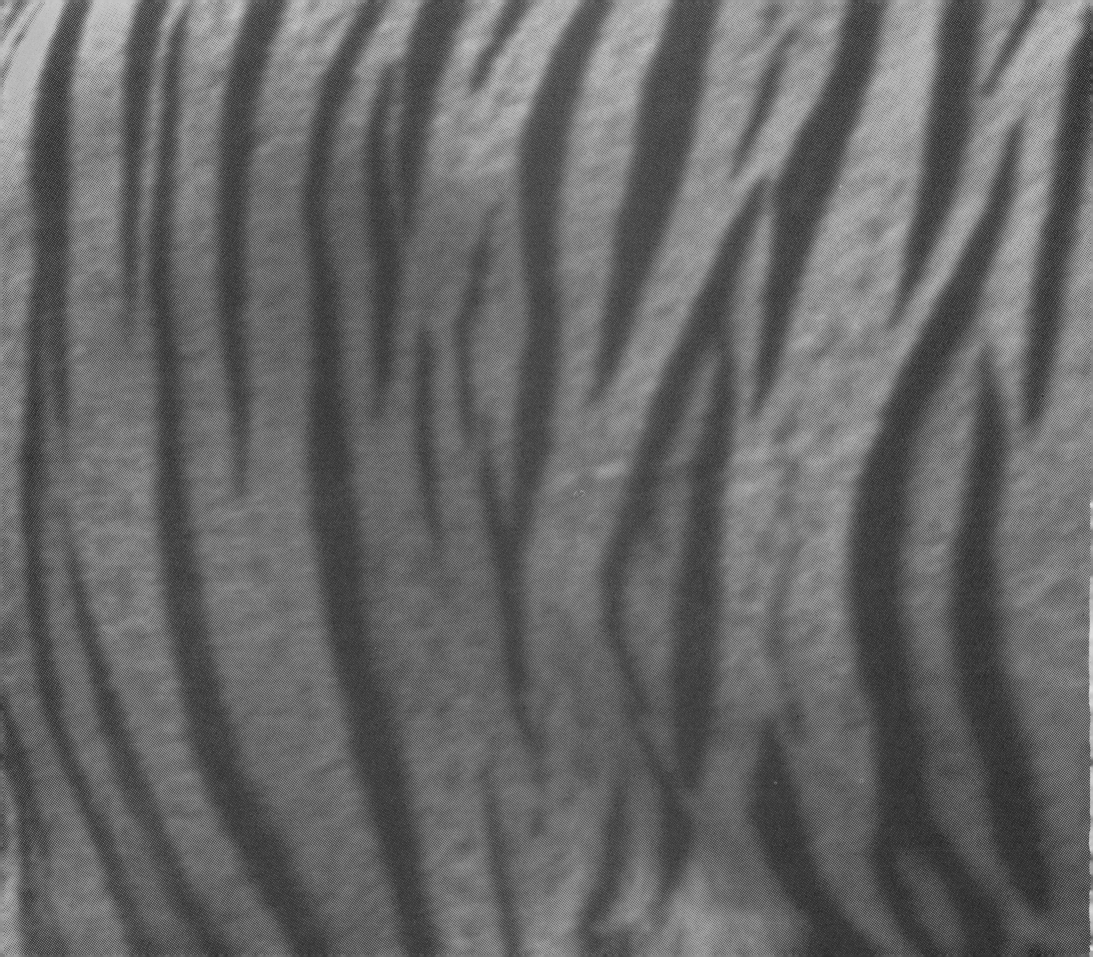

R.G.S. GUILDFORD.
J. C. MALLISON
LIBRARY.

A BEVY OF BEASTS

ACC. No.	CLASS No.
234239	591
1/98	

To my friend
and advocate:
George Medley

Edward ((Lendt ((
June 1993

R.G.S. GUILDFORD,
J. C. MALLISON
LIBRARY.

EDWARD MENDELL

A BEVY OF BEASTS

THE ENCHANTING ANIMALS OF BORNEO, BELIZE & BEYOND

WITH FOREWORD BY

HRH PRINCE PHILIP

DUKE OF EDINBURGH KG KT

Although all the

animals in this book

are unusual, they

are arranged in

order of their degree

of endangerment,

starting with those

most at risk.

This book is dedicated to my grandchildren

■ A M Y   ■ J E R O M E
■ T O N I   ■ L I N C O L N
■ J O D I   ■ P R E S T O N
■ R O N A L D ■ A L Y C I A
■ E L Y S E   ■ M E L A K U

with the hope that when they, and their grandchildren, go on their own safaris, they will find all the species shown here still roaming free in their natural habitats.

Finding the wild creatures that inhabit the far-flung corners of our planet is more than a journey of discovery... it is an awe-inspiring look into nature's fabric.

I have been privileged to make these journeys and to be able to share some of these wonders with those who read this book.

Along the way I have always been encouraged by Renée, my life's partner, and I am eternally grateful for her unflagging support and counsel.

Copyright 1993 by Edward Mendell

Library of Congress Catalog Card Number: 92-96981

ISBN 0-9515863-1-9

British Library Cataloguing in Publication Data
Mendell. Edward
A Bevy of Beasts
1. Animals
I. Title
591-9598
ISBN 0-9515863-1-9

CONTENTS

A NOTE FROM THE AUTHOR

Finding and photographing wild animals is rarely easy and may oftentimes be difficult and demanding. Difficult terrain, uncooperative weather, hard-to-find spoor, wary animals' apprehensiveness, these and many other things can make the task foreboding. But it is a quest that promises always to be exciting.

When I am on safari, at the end of each day I enter notes into my travel journal - a sort of diary - so that in the future I can recollect the happenings and impressions of that day.

While capturing every animal image in this book was a fascinating adventure, some were especially memorable and for these I've included both the notes from my diaries and additional photographs

———————

All the photographs in this book were made with a Hasselblad camera using Carl Zeiss Hasselblad lenses, either F2/110 mm, F4/250 mm or F4/350 mm.

All the images were made on Fujichrome 100-D film.

A first book, WILDLIFE ODYSSEY, published three years earlier, is the companion volume to A BEVY OF BEASTS. In WILDLIFE ODYSSEY the author chooses nine sites which he believes offer the best opportunities to observe fascinating wildlife. He comments on the regions and presents an array of his photographs to illustrate the allure and variety to be found in each.

All profits from the sales of that book, and of this book, will be donated to the World Wildlife Fund/World Wide Fund for Nature.

BUCKINGHAM PALACE.

One of the most important functions of any
conservation organisation is to make more people aware
of what is happening to the natural environment. With
such a high, and growing, proportion of the population
living in towns and cities, fewer and fewer people are
directly in touch with the natural world. Even people
who live in the country are more likely to know about
agriculture than about the wild species of plants and
animals. That is why books like this one are so
valuable.

The author has put together a splendid collection
of evocative photographs and he has a written a clear
and unequivocal text about the threats to the survival
of so many magnificent beasts. I hope it will stir the
reader's conscience and persuade more people to take a
more positive attitude to the conservation of nature.

INTRODUCTION

`FACT` Although scientists have estimated that the earth may contain more than 25 million living species, they have identified and named only 1.4 million.

`FACT` Those identified include 22,000 fish, 10,500 reptiles and amphibians, 9,000 birds and 4,000 mammals.

`FACT` Of the 325,000 plants that have been classified, only 5,000 have been analyzed for potential medicinal value.

`FACT` Of the mammals, humans are by far the most numerous. Mankind now numbers in excess of 5 billion individuals.

`FACT` We humans are currently annihilating one living species every hour of every day. That's equivalent to more than ten thousand species per year.

Two examples of this extirpation are the elephant bird in Madagascar and the moa in New Zealand. These were huge birds - the egg of the elephant bird had a two gallon capacity; the moa grew to a height of 15 feet. Both were annihilated by man. And when a species becomes extinct it is gone forever - those who have caused it, by whatever means, have despoiled the earth for all the generations to come.

The earth is now at its lowest biodiversity since the Mesozoic Era over six million years ago.

Kenya's elephant population has dropped from over 60,000 to less than 15,000 in only fifteen years; in that same period the number of rhinos has been disastrously slashed from 15,000 to a mere 500!

Although 54 countries have set aside protected areas in the attempt to conserve wildlife, the human threats continue. These include mining, logging, expansion of agriculture and grazing, acid rain, water diversion, unconstrained tourism, etc.

Unless the wanton destruction is stopped it is probable that, in only a few human generations, the scenes in this book will be impossible to photograph because the pictured animals will no longer roam these habitats.

The collection of photographs in this book was assembled with the hope that, as others see the wonder and beauty of these fascinating creatures, they will support those activities which are aimed at conserving animal species - because the happy fact is that there are a number of worthy organizations and dedicated people doing something about protecting wildlife.

The animals in this book are, for the most part, rarely seen by the traveller journeying to the better-known wildlife areas. But in addition to their scarcity, many of the species are also highly endangered. Thirty-seven of them are at risk, and many of those may soon disappear forever without drastic measures to ensure their survival.

PART ONE

FACING IMMINENT DANGER OF EXTINCTION

WILD DOG
OR
HUNTING
DOG
Lycaon pictus

THIS is a fairly large carnivore having a body similar in appearance to a big domestic dog, but with a massive head almost like a hyena, thin legs, large ears and mottled coloring. Its feet have only four toes.

Living in packs of up to fifty individuals, it is a most efficient predator, attacking prey of all sizes, from elands and zebras to small hares. Hunting usually in the early morning or late evening, a wild dog can run at speeds up to 36 miles an hour. It will cooperate with other wild dogs in the hunt, following slowly before beginning the final chase and, as the lead dogs tire, taking over the chase. When close enough, the wild dog starts to bite whenever it can, ripping the animal into shreds with its sharp teeth and often seizing its legs and tail, thus causing it to fall.

It will then immediately start tearing it to pieces and feeding, sometimes while the prey is still alive. Adult wild dogs share the kill without aggression and very often allow the young dogs to eat first. The entire prey may be consumed within a few minutes.

A wild dog does not drink from water holes, getting enough liquid from the blood and body fluids of its prey.

▨ Wild dogs, such
as these in the
Transvaal, have
excellent sight,
hearing and scent.

ASIATIC ELEPHANT
OR
INDIAN ELEPHANT
Elephas Maximus

THE Asiatic elephant has smaller ears than its African counterpart and a more humped back as well as a more concave forehead. The female has only rudimentary tusks. The family is usually composed of ten to thirty individuals led by an old female and includes several females, their young and an old male, all of which are usually related.

The herd rests during the heat of the day, spending its other daylight hours feeding on grass, leaves, shoots, fruit and other plant materials, all of which are grasped using the elephant's highly sensitive trunk. While hearing and scent are excellent, an elephant's eyesight is poor. When walking, the herd usually travels in single file but if there is any danger the adults surround the young. If a member of the herd is shot or wounded, others stay to assist it, often putting themselves in great danger.

In densely wooded areas, where an elephant might lose sight of the others, it emits a low, grumbling purr to maintain contact. Whereas the African elephant lives for up to fifty years, Asiatic elephants live for as long as seventy years. The African elephant has at the tip of its trunk an upper and lower triangular projection which can be used for gripping but the Asiatic elephant has only one such projection.

Asiatic elephants, like these in Assam, live at elevations up to 12,000 feet.

GREAT INDIAN RHINOCEROS

Rhinoceros unicornis

. . . .

THE most remarkable feature of the single-horned great Indian rhinoceros, the largest of the Asian species, is the thickness of its skin which, with its deep folds, gives the rhinoceros an armor-plated appearance. The hide is covered with many small protuberances which are scale-like in appearance. The feet are all three-toed and the legs are short and powerful.

Generally a solitary animal while eating, the Indian rhinoceros shares wallowing and drinking places. It feeds in the morning and evening on grass, weeds and twigs and rests during the hotter mid-day period.

The Indian rhinoceros mother is accompanied by her single offspring for several years. It rarely attacks, even in defense of its young, and usually tries to flee from possible danger.

The great Indian rhinoceros likes to be near water and bathes frequently.

For centuries people in the Orient have believed that many parts of the great Indian rhinoceros have magical properties and as a result the species has been hunted almost to extinction.

Today, the great Indian rhinoceros population is numbered only in the hundreds.

March, 1992 - Kaziranga, Assam, India

There are less than 400 great Indian rhinoceroses in the world and I'd come here hoping to see them in their natural habitat.

Since the grass is taller than a man and the wet earth impassable because of the elephant tracks that potmark every foot of the ground, the only way to move about is on the back of an elephant.

Each morning I met my mahout and his elephant and we clumped through the vegetation in search of the wary rhinos. Each time we attempted to approach one, it turned tail and fled.

On the second day we discovered a waterhole and found a rhino drinking. We were finally able to come close enough to a rhino to capture some memorable images.

And the next few days were just as successful as, on each of them, we were able to approach more of the enthralling one-horned rhinos, observing them from our jostling perch on the elephant.

GOLDEN LION TAMARIN
Leontopithecus rosalia

. . . .

*T*HIS beautiful monkey is a particularly small primate, having a body length of only seven or eight inches. The silky golden mane which covers its head and shoulders gives this New World monkey its name. It is an agile animal, leaping from limb to limb in its search for food: insects, fruit and birds' eggs.

Primarily a tree-dweller, the tamarin does not have grasping hands or the opposable thumbs which distinguish most primates. It jumps from tree to tree in squirrel-like fashion but does not swing on vines or branches. It rarely comes to ground, usually remaining close to the very tops of trees.

It is diurnal, constantly on the move during the day, spending its night in holes in the trees. The golden lion tamarin lives in small groups of four or five adults. It is a vocal creature, issuing high, shrill squeaks.

Both parents help care for their young but it is the father who offers them their first solid food, after pulverizing it with his fingers.

■ The golden lion tamarin is found only in the Poco das Antas Reserve in Brazil.

EASTERN LOWLAND GORILLA
Gorilla gorilla graueri

THE gorilla is the largest and heaviest of the primates. Living mainly on the ground, an adult will venture into the trees only to build nests for the night. An older male usually sleeps in nests on the ground. However, a young gorilla spends much time playing in the trees.

Moving on the ground in single file when changing locations, the gorilla walks on all fours in a kind of knuckle-walking. An old male will stand erect when chest beating but bipedal movement is uncommon. The gorilla's sight and hearing are very good and its scent is exceptional.

It is a vegetarian, eating leaves, buds, shoots, stalks, roots, tubers, bark, ferns, etc. The diet includes over one hundred distinct species of food plants.

Fighting among older males begins with a yawn and a direct stare since, as face to face vision is always avoided, that constitutes a challenge. When confronting strange animals or men, a gorilla will often bark and roar and beat on its chest. Nevertheless, a gorilla is a peaceful creature that will only attack if threatened.

■ The eastern lowland gorilla inhabits the mountain range on the Zaire-Rwanda border.

22

June, 1992 - Kahuzi-Biega, Zaire

Accompanied by an armed guard, a porter and two trackers, I climbed high into the mountain's thick, dense brush. Following the machete-wielding trackers, we followed the day-old spoor of a gorilla family.

After five hours of exhausting climb one of the trackers pointed excitedly to a tangled clump of vegetation. I crept close and parted the foliage - there was my first gorilla!

An adult female, she stared back, but never stopped munching on the leaves of a long vine. When she ambled off I followed and she led me to a small clearing where several youngsters were wrestling, climbing trees and tumbling all over the ground.

One of them, bolder than the others, bounded over and sat directly in front of me. What a thrill.

The next day I met up with a silver-back adult male who rose up, stomped to within a few yards of my position, beat his chest, barked hoarsely, turned and walked away.

What emotions I experienced with these "fellow primates."

Eastern lowland gorillas inhabit both bamboo forests and mountain rain forests.

TIGER
Panthera tigris

■ Considered a threat to other living species, it is doubtful that a tiger could survive outside a reserve area such as this in Ranthambhor, India.

THIS largest of all the cats, a shy, unsociable animal that remains deeply hidden in dense brush during the day, is also the most difficult to spot in its natural habitat.

It has a massive, muscular body and powerful limbs. Its coat differs with its geographical habitat; the Bengal tiger is relatively short-haired and has strongly marked stripes, the Siberian tiger has longer hair and fewer, paler stripes. In the shadows of a forest, its stripes help a tiger to disappear from view.

Remaining hidden during the day, it stalks its prey only after dark. In hunting, it often lies in ambush, then gallops in pursuit of whatever is available, mainly medium to large mammals such as cattle, wild boar, deer, buffalo and, as well, man, being one of the few animals that regularly preys on humans. Prey that is too large to be eaten at a single feeding is dragged away and hidden in a thicket to which the tiger will return on the following days.

The tiger maintains a clearly staked-out territory which it patrols at regular intervals. It climbs well, swims well, moves gracefully on land, and can attain great speed over short distances when chasing prey.

Capable of killing animals twice its size, the tiger is one of nature's most feared predators.

HYACINTH MACAW
Anodorhynchus hyacinthinus

Of all the parrots, this brilliantly plumaged one is the world's largest. However, unlike most other large parrots, which fly slowly and laboriously, the hyacinth macaw flies swiftly, frequently almost touching the wings of others in the group. When seen, it is usually one of a group of fifteen or twenty, either flying majestically over forested areas or perching among the branches of a palm tree.

To get to its main food, nuts, it climbs palm trees, using both its feet and beak. Its nesting sites are in the hollows of these and other trees. It is a particularly noisy bird, issuing harsh squawking shrieks while in flight.

The hyacinth macaw has a very long life for birds, some of them living to an age of sixty or seventy years.

The destruction of habitat by farmers and ranchers as well as the capture of young hyacinth macaws for the illegal bird trade make this a highly endangered species.

■ The brilliantly colored hyacinth macaw is one of the rarest of all parrots.

■ A newborn white
lion is snow-white
but then becomes
light honey colored as
it matures.

*T*HE largest of the African cats is the lion, normally ranging in color from reddish-grey to a pale tawny with lighter colored underparts. The back of the ears is black. The adult male often has a mane of long hair which ranges, geographically, from tawny to black.

In South Africa's Transvaal Lowveld, in the region of Timbavati, there have been sightings of "white" lions, the first truly white lions the world has ever known. They are not albinos, but genetic variants with white pigmentation.

There have been a few known instances of true albino lions that appear to be white because of lack of pigmentation. This also affects the color of the eyes.

The white lions of Timbavati, whose mothers are rarely white themselves, are positively white-pigmented, with their normal yellow eyes being only a shade lighter than their mother's.

WHITE LION
Panthera leo

■ This white lion's
behavior and habits
are no different from
those of her tawny
colored sibling.

June, 1992 - Timbavati, South Africa

Sitting around the campfire at Motswari, talk was stimulated by the repeated sighting of the rare (only 2 living examples in the wild) white lions.

Before sun-up my guide and I, already on the road, found tracks of a young giraffe in flight and a pursuing lion.

We followed the trail into dense jungle where, obviously unable to keep up, the lion's prints veered off. We criss-crossed the area for hours until we located the pride in a dense thicket.

Peering through the verdant growth I could see a handsome, golden-brown cub. And then, in a matter of minutes, that tawny lion moved away to reveal two other cubs, one of which was a <u>white lion</u>. I could hardly believe my good fortune and quickly photo-graphed her.

Later in the day we spotted the other, older white lion, nonchalantly strolling across a dirt road right in front of us.

In all the world there are only two white lions in the wild. I had found and photographed both of them.

WILD ASIATIC WATER BUFFALO
Bubalus bubalis

THIS large, thickset, clumsy creature with a long, narrow face, has the largest horn span of any living cattle. Feeding early and late in the day, the water buffalo eats the lush grasses and other vegetation that grows along river banks and lake shores. During much of the rest of the day it spends its time submerged, with only its muzzle showing above water, or swimming in marshes and swamps. It frequently wallows in order to cake itself with mud as a defense against insect bites.

The water buffalo is a gregarious animal that lives in herds of up to twenty individuals. Tame and docile, it has been domesticated and used as a beast of burden for at least 5000 years. Additionally, it yields milk of good quality and its hide makes excellent leather.

The water buffalo's main predator is the tiger, but it rarely succeeds in killing a buffalo because the herd bunches together with the powerful horns turned outward in defense.

▨ Both male and
female wild Asiatic
buffalo have horns.

SOUTHERN RIGHT WHALE
Eubalaena australis

WEIGHING about 60 tons, about 45 feet in length, the southern right whale is easily identified by the skin protuberances (callosities) on its head just in front of the blowhole. These protrusions are infested with barnacles and parasitic crustaceans. Another identifying characteristic is its deeply curved jaw line.

It was named "right whale" because, as a slow swimmer, almost always afloat when killed, it was the "right whale" to hunt and destroy. Furthermore, it had a particularly high yield of baleen and oil, both products of great commercial value and, in addition, its blubber could be up to 16 inches thick.

The right whale has tremendous food needs, devouring about five thousand pounds of food every day as it feeds by skimming through surface concentrations of zooplankton with its huge mouth opened. Its normal dive duration is up to fifteen or twenty minutes.

Although the right whale usually leads a solitary life, small groups of up to six may stay together for short periods of time.

■ Right wales such as these in Patagonia have been hunted ever since the ninth century.

ORANGUTAN
Pongo pygmaeus

THIS peaceful, gentle, long-haired primate lives in the rainy jungles of Borneo and Sumatra. Clumsy when moving on the ground, it spends most of its time high up in the trees. At night it sleeps on little platforms which it builds from twigs and branches, constructing a new one each night. In wet or cold weather the orangutan covers itself with leaves. A voracious eater, it will sometimes spend an entire day feasting upon the fruit of a single tree. Fruit makes up more than half of its diet but it will also eat bark, leaves, eggs and young birds.

Long arms, hook-shaped hands and feet, and a heavy body enable an orangutan to swing superbly from tree to tree and walk on the branches. A very young orangutan rides about clinging to its mother's chest but juveniles often play and romp together. Adults are quite solitary and, other than during infrequent sexual encounters, an orangutan maintains and protects its own independent home range. Even when several adults are eating at the same food source there is no evident interaction among them. An adult male is particularly jealous of his territory and, when encroachment does occur, will become quite aggressive and noisy, shaking branches and emitting long, loud howls.

■ The most arboreal of all the apes, the orangutan has especially long, flexible fingers that help it to hang and swing in the trees.

■ The orangutan has
wonderful ability to
use its facial muscles,
particularly its big
lips.

May, 1991 — Sabah, East Malaysia

I had come to this rain forest in northern Borneo to photograph orangutans (from the Malay word meaning "man of the forest").

At mid-day we sat down on a rotting log, leaning my brightly colored umbrella at the far end. Suddenly there was a "whoosh" sound and I watched in amazement as a young orangutan swung down on the vines of a nearby tree, snatched the umbrella, and swept back into the high branches.

Within seconds a small crowd of other chattering orangutans joined him and my umbrella was soon in shreds. The inquisitive youngsters kept swinging close by, trying, with arms outstretched, to grab my camera and tripod.

There was soon a crowd of orangutans, all curious, including even a female clutching her small infant.

As we moved along through the overhanging limbs and branches, the orangutans moved right along with us, giving me many opportunities to make exciting photographs.

PART TWO

DEPLETED AND HIGHLY VULNERABLE

GIANT ANTEATER

Myrmecophaga tridactyla

THE giant anteater is the largest of the species. It has no teeth but its tongue is covered with sticky saliva which it uses to catch insects, particularly termites and ants and their larvae. The giant anteater is always on the ground unlike other members of the anteater family, all of which are able to climb and usually dwell in trees.

In addition to using its very sharp, powerful claws to tear into termite mounds and ant hills to get at the insects, larvae and eggs, it uses them as well to defend itself from its main predator, jaguars, and its claws can be especially dangerous. To protect them, the giant anteater walks on its knuckles.

Although its sight and hearing are poor, the giant anteater has an excellent sense of smell, forty times more powerful than a human's.

It sleeps for up to fifteen hours a day. At night it beds down in the burrow of another animal or scrapes a hollow in the soil.

■ A giant anteater is easily recognized by its long snout and long, bushy tail.

A single, young giant anteater is carried on its mother's back and will remain with her until the next one is born.

September, 1991 – Das Emas, Brazil

Deep inside the Mato Grosso wilderness I was hunting for the rare giant ant-eater. Catching sight of them is easy. Getting close enough to photograph one is another thing. Their keen sense of smell detected my presence long before I could come within camera range.

Late in the day I spied a magnificent adult male and, realizing the only chance I had was to approach from downwind I entered the bushy pampas well down from where he was roaming. Even though I knew he had poor eyesight I kept hidden behind foliage as I approached.

Suddenly he came into view, trod along unconcerned, then abruptly turned and walked towards me.

I took several photographs, then stood stock-still as he passed within a few feet of where I was standing.

Sensing a foreign presence he raised his snout high in the air and waved it about. Once past me, he changed direction once again and scurried away.

KING CHEETAH
*Acinonyx
jubatus (rex)*

. . . .

THE cheetah is the swiftest animal in the world, able to run at speeds up to 70 miles an hour. Living in open territories, using its excellent sight, it hunts for gazelles and antelopes both in daylight and periods of full moon. It catches its prey by rushing at them with great speed and pouncing on them.

Cheetahs have a rather rough, tawny-colored coat that is entirely covered with small, solid, round, black spots, scattered singly, unlike those of the leopard which are in rosettes. The cheetah's spots are lighter on the underparts, becoming almost white on the belly.

However, the "king cheetah", a rare intraspecific variation found only in a small area of Zimbabwe, is characterized by long, broad stripes, very irregular in length and shape, longitudinal on the back and diagonal on the flanks.

It is not, in fact, a different species, but an individual variant of the normal type.

■ The cheetah has been tamed for hunting for almost 5000 years.

OCELOT
Felis pardalis

$\mathcal{T}$HE ocelot may be found living in a wide range of habitats, from semi-desert to tropical rain forest. Although coloration may differ depending on the area in which it lives, all ocelots have spots on the head and limbs, stripes on the face and insides of forelegs. It normally sleeps during the day, covered by branches or heavy vegetation, traveling along the ground at night in its search of prey. Extremely secretive, it rarely shows itself in open country.

In its hunt for food it lies well-hidden behind bushes or branches while waiting for small deer, monkeys, rodents, snakes or other reptiles which it then pounces upon. It will also eat domestic animals up to the size of a calf. Although it has a highly developed sense of smell, it relies more on its hearing and eyesight to detect prey on its nightly hunting expeditions.

Living in pairs, in well-defined territories, each individual ocelot hunts alone but the pair may work together to kill a larger animal. An ocelot will eat any animal it can overpower.

When ocelots mate, always at night, courting males make loud, screeching sounds, not unlike those of domestic cats.

An ocelot, like this one in Belize, was severely exploited as the skins of hundreds of thousands were used in the making of fur garments.

KOMODO DRAGON
Varanus komodoensis

THE awe-inspiring Komodo dragon, discovered only in 1912, is the largest of the monitor lizards and dwarfs all other present-day lizards. It has a powerful body, long, thick tail, well developed limbs with talon-like claws, large, jagged teeth and a forked tongue that can be flicked in and out of its mouth. It uses its tongue as one of its sense organs, constantly flicking it about as it moves along the ground.

Despite its size it can swim well, climb well and move rapidly over land. Active during daylight hours, it preys on animals as large as hog deer and wild boar, as well as on small deer and pigs. It also occasionally attacks wild horses. The Komodo dragon is so voracious that it will attack and devour its own kind and will feed on the dead bodies of all other animals.

Living only on two small Indonesian islands, the Komodo dragon is considered a relic of earth's giant reptiles long since extinct and, as such, is of great scientific interest.

■ To conserve the small remaining number of Komodo dragons, the island on which they are found has been declared a nature reserve.

JAGUAR
Panthera onca

*T*HE only member of the genus Panthera (big cats) in the New World is the jaguar, the largest, fiercest carnivore of South America. Unlike the big cats of Africa, it does not roar; instead, when threatened, it emits a loud snarl. Normally living in thickly wooded areas and grassy marshes the jaguar will, when necessary, pursue game over open areas. Other than during the few weeks of the breeding season, it is a solitary animal, constantly on the move, and frequently traveling enormous distances - as far as five hundred miles. It hunts mainly at night, preying on both aquatic and terrestrial animals.

Like leopards (to which it is often compared) the jaguar climbs trees to wait for prey. It is an excellent swimmer. Jaguars eat most of the smaller wildlife including peccaries, capybaras, turtles, otters, sheep, monkeys and terrestrial birds. Its name is derived from the South American Indian word, "jaguara," which can be translated, "carnivore that captures its prey in a single bound."

It is widely believed that the jaguar catches fish by flicking its tail on the surface of the water as a lure and then, using its forepaw, tossing the fish up onto the bank.

Newborn jaguars are blind at birth and do not open their eyes until they are about two weeks old. A young jaguar will remain with its mother for up to two years.

■ A jaguar such as this one in Brazil will fiercely defend its urine-marked territory.

JAGUARUNDI
Felis
yagouaroundi
....

A slender, weasel-like cat, the jaguarundi, with its short legs, long tail and small, flattened head is found in dense, bushy, lowland forests where it hides in low branches and brambles. It is a solitary animal, pairing only for breeding.

Although primarily terrestrial, it moves about with considerable agility on both the ground and in trees in its search for food, mainly birds and small mammals such as rats, mice and rabbits.

In two of its habits, it is unlike most other cats: it remains on the move throughout both night and day and, being an excellent runner, it sprints after and overtakes even the fastest prey, unlike the usual cat habit of stalking and ambushing prey.

■ A jaguarundi, like this one in Brazil, is far more active during daylight hours than most other cats.

MARGAY
Felis wiedii
. . . .

Similar to the ocelot, although somewhat smaller and slimmer, the margay is also known as the "tree ocelot" or "American tiger cat." This long-tailed animal, one of the smaller South American cats, has been hunted so extensively for the pet trade and fur industries that it is now found only in some of the densest forests of Mexico and Central America.

It is a nocturnal animal, feeding mainly on birds and small mammals. Living most of its life in trees, the margay is a particularly skillful climber although, unlike most other cats, it descends from trees headfirst. Its hind feet are unusually long-clawed which allows the margay to hang from a branch using only one hind foot.

■ This margay, in Belize, having just captured its prey, is resting after the pursuit.

BLACK SPIDER MONKEY

Ateles

paniscus

THE spider monkey is a monkey with arms longer than its legs and a very long, strongly prehensile tail. The generic name, Ateles, means "imperfect" and refers to the absence of any thumbs. Yet, with four fingers all working in the same direction, it is an extraordinary climber, surpassed only by gibbons for agility in the trees. Its tail is the most highly developed of all mammals and is used like a fifth hand, permitting the monkey to hold on to branches while exploring for fruit and nuts. It is when hanging by its tail, with all other limbs free, that it resembles a spider.

Living in trees in groups of fifteen or twenty, the spider monkey rarely comes to ground. Most feeding is accomplished in the early morning and late afternoon. Having no thumbs, the monkey often uses its highly sensitive tail to gather and hold food such as fruit while it takes off the skin with its teeth. Members of the troop use long, loud calls to coordinate movements.

The spider monkey is able to travel through the trees at amazing speed, moving in a style called brachiating: progressing from one branch or vine to another by alternate grips, using its hands and tail in the process.

■ This Brazilian spider monkey's tail is strong enough to hang by yet, using the hairless undersurface at its tip, is sensitive enough to probe for tiny nuts.

CHIMPANZEE
Pan troglodytes

*P*ROBABLY the most intelligent animal, next to man, the chimpanzee manifests many parallels to human behavior. Although its senses are similar to ours, it probably has a keener sense of smell. Other than man, the chimpanzee is the only animal that makes and uses tools. For example, it peels the bark from sticks and then uses those bare sticks to probe tree holes for insects. A chimpanzee uses far more facial and body gestures and a wider range of sounds than any animal other than man.

Usually in a slouch, rarely holding itself upright, a chimpanzee walks on its knuckles while moving about mainly on the ground, although it also climbs. A chimpanzee spends many hours of each day foraging for food, mainly ripe fruit and young leaves. It may eat as many as two dozen different types of plants in a single day. A chimpanzee will also eat small animals, insects, young pigs, antelopes and monkeys.

A newborn chimpanzee is totally dependent on its mother. When a few days old it clings to the front of its mother's body, by six months it rides on her back and by four years is walking alongside. It will then remain close to its mother for up to eight years. Chimpanzees live in communities ranging in size from about a dozen to over one hundred.

■ Like other arboreal monkeys. the chimpanzee is very noisy, issuing hoots and high-pitched screams at the slightest provocation.

Trying to get my attention, this young chimpanzee hooted noisily while, at the same time, making the most amusing faces.

September, 1990 – Gombe, Tanzania

These remote forested mountains are
one of the few remaining sites still
inhabited by chimpanzees in the wild.

Isolated from civilization, with no way
of communicating with the "mainland",
other than my guide, I was the only human.

We clambered up mountains, struggling
ever higher in the attempt to reach
the spot where we heard the
chimpanzees chattering – but had
no sight of them all that day.

While walking through the forest
the next morning, I was startled
when a big adult male chimpanzee
bounded out of the bush, stopped just
in front of me and glared fiercely.

Immediately I sat down, lowered my
gaze, and waited. Mollified by my
submissiveness, he signalled with a
grunt and his female and youngster
entered the clearing as well!

I remained with them for several
hours as they traipsed through the
jungle, exhilarated by being this
close to them.

CARACAL
Felis caracal
. . . .

HIS solitary animal, identified by its long, slender legs, big paws, flattened head and long, tufted ears, is the lynx of Africa. It is a nocturnal cat that often lives in hilly country where it hides among rocks in well-marked territory during daylight hours, avoiding the heat. It hides its young in crevices or hollow trees. It is an excellent climber.

The caracal feeds on various small mammals such as young antelopes, rodents, hares and lizards. Among the reptiles it eats are poisonous snakes which it takes by surprise. It is often destructive to domestic sheep, goats and poultry.

It will also eat birds which it catches by springing high up into the air to catch birds starting off in flight. So great is its agility in this that Eastern princes used to use caracals in pigeon-catching contests.

The caracal's senses of sight and hearing are quite good but its sense of smell is only moderate. In former times the caracal was trained for hunting in India.

■ In pursuit of its prey, a small caracal like this one in Samburu can spring into the air up to six feet.

SWAMP DEER
Cervus duvauceli

. . . .

*T*HIS animal, also known as a barasingha, varies in color from brownish to yellowish. It is most often seen with small white spots on its back and side. In order to ensure better footing in the wet marshy area in which it lives, the swamp deer's hooves are large and splayed.

Although it is an extremely aquatic animal, almost always found close to marshes or grassy areas near water, it may also inhabit forested regions. It is a grazing animal, eating in early morning and late afternoon, resting during the remainder of the day in cool, wet places.

When threatened by any of its usual enemies such as tigers or leopards, the swamp deer will instinctively head for water, quickly entering and swimming to safety.

■ The swamp deer lives in flat areas that are moist and humid as in Kaziranga.

ELAND
Taurotragus oryx

*A*very large antelope with a bovine appearance, the eland has massive horns which lie backwards in line with its face. The sides of its body have light stripes.

Usually living in herds of up to about seventy head, led by one or two mature bulls, it often mingles with roan and zebra. For such a heavy animal it is a superb jumper, able to leap over other elands from a standing start.

It feeds on leaves, bushes and several varieties of fruit as well as large bulbs which it digs for with its hooves. The eland uses its horns to collect small branches, grasping and twisting them off with a movement of the neck. Although it drinks regularly when water is available, it can go for long periods without drinking.

Although its sight is relatively poor, its hearing and scent are excellent. Being a non-aggressive animal, easily tamed, and more quickly fattened than any other African antelope, the eland has been kept in semi-captivity for its flesh and its milk.

■ Preyed upon by lions and wild dogs, the eland mother will bravely defend her young. This one is in the Transvaal.

COMMON LANGUR
Presbytis entellus

THIS black-faced, large, long-limbed monkey is a very adaptable creature that is most frequently found near human habitation in the outskirts of villages where it scavenges for food. Its body is silver-haired with black face, hands, ears and soles of feet. Its face is fringed by grey-white whiskers.

Although an excellent climber, the langur spends as much time on the ground as it does in the trees. It is a vegetarian, consuming leaves, seeds, buds and fruits which it finds in the foliage.

Although protected in some areas, and considered sacred by Hindus, langurs are often ill-treated and hunted in some regions where they cause considerable damage to agricultural crops.

A newborn infant clings closely to its mother for up to four weeks after which it begins moving short distances on its own.

■ Although langurs like this one in Ranthambhor are organized in a dominance hierarchy, a female that has just given birth moves out of this structure.

■ Suspicion among langurs is such that a male, leaving his own group, is seldom accepted by another.

BLACK HOWLER MONKEY
Alouatta caraya

THE largest of the New World monkeys is the black howler, the adult male being the size of a fairly large dog. The black howler produces unbelievably loud shouts as a result of having a special bulbous larynx which creates tremendous resonance. The calls, which can be heard for miles, are probably used to define and defend the black howler's territory. Most of the shouting is done in the early morning and late afternoon.

A powerful monkey with strong limbs and a prehensile tail, it lives most of its life in the trees where it feeds on leaves and fruit, especially figs. It can make remarkable leaps, using its tail for support. Only the male adult is black. The female and young are pale yellow.

The female gives birth to one young and her offspring will remain with her for up to two years.

■ A black howler will roar loudly whenever it moves from one feeding site to another.

PACA
Cuniculus paca
....

A solitary, nocturnal rodent, the paca spends its days in burrows which it digs along river banks, among tree roots or under rocks, emerging after dark to hunt for food. Frequently found close to water, the paca is a good swimmer and, if threatened, will retreat to the water where it can stay submerged for a remarkably long time. If not near water when faced by a predator, a paca leaps away and "freezes" in its landing position, unmoving from this pose for as long as 45 minutes. When walking in the forest, it makes use of worn trails.

The paca has a handsome brown coat with horizontal rows of creamy spots down the sides. Its most unusual physical trait is the composition of its back which consists of a peculiarly fragile skin that slips and slides over a thick connective tissue, thus making it very difficult to grasp.

The paca feeds on a variety of vegetable matter including roots, leaves, seeds and fallen fruit.

■ In some areas of Brazil the flesh of a paca is in great demand, thus fetching very high prices.

SOUTH AMERICAN SEA LION
Otaria flavescens

THE South American sea lion is an eared seal with a broad upturned muzzle. Unlike true seals which use their flippers for swimming, the sea lion makes long sweeps with its foreflippers but the hindflippers have no role in sustained swimming. On land the sea lion is much more agile than the true seal - whereas the true seal humps along on its belly, the sea lion moves with its body clear of the ground as both foreflippers and hindflippers effect a galloping kind of locomotion.

An adult male South American sea lion is about eight feet in length and weighs about 650 pounds. The adult female is two feet shorter and weighs around 300 pounds. Both sexes are dark brown, but newborn pups are almost entirely black.

The average harem consists of a bull with nine or ten cows. The harem will remain intact until the pupping and mating times are completed or, in some instances, when the bull loses a battle with an invading bachelor. When frightened a sea lion will always take to the water.

■ Although protected from hunting, there is still great damage to the South American sea lions caused by Argentine fishermen's nets.

GRIZZLY BEAR
Ursus arctos

A large, immensely strong brown bear, the grizzly is among the biggest carnivores. Generally moving with a low, clumsy walk, swinging its head from side to side, the grizzly can, when necessary, move as fast as a horse. Its paws are broad and flat with long non-retractile claws. One blow with this paw can kill an animal of equal size.

It feeds on berries, nuts, roots and aquatic plants, but its preference is for the flesh of fish or mammals. It can scent carrion from as far away as twenty miles. In the autumn, when salmon migrate upstream to spawn, the normally solitary bear will join with other grizzlies that congregate along rivers.

Although a grizzly will normally avoid any contact with man, it is the most unpredictable and dangerous of all bears. In its natural habitat it is king, has no enemies or predators, and very rarely fights with other animals.

Primarily nocturnal, a grizzly normally beds down in the daytime in hidden thickets interlaced with leaves, small branches and needles.

It is the silver-tipped hairs on its back and shoulders that give it the "grizzled" appearance from which it gets its name.

■ Although there are considerable variations in the size of grizzly bears, by far the largest are in Alaska.

August 1991 - Katmai, Alaska

Every autumn more than a million salmon
burst from the Bering Sea and enter the
rivers of this tiny island. During those few
weeks Alaska's huge grizzly bears also
come to Katmai - to feed on the spawning
salmon.

I came here by float plane. I wanted
to see and photograph this spectacle
at the place where a waterfall
spans the Brooks River - a place
where the most powerful bears stand
at the top of the falls and catch
the jumping salmon with a quick
snap of the jaws.

When a salmon is caught, the bear may
eat it on his rocky perch or, more often,
walk ashore with his prey and eat in
on dry ground.

The bears were so intent on catching
fish that, as I had anticipated, it
was possible to get much nearer to
them than at any other time and
I could capture remarkable close-ups.

PART FOUR

Of Conservation Concern

HONEY BADGER OR RATEL

Mellivora capensis

*L*IVING alone or in pairs, the honey badger is a brave nocturnal animal that will not hesitate to attack creatures much larger than itself. With a thickset, compact body, large head and powerful claws, it lives in large burrows in the ground. It is easily identified by the black underside which contrasts with the lighter colored back.

In addition to animals, it feeds on plants and insects, particularly honey and bee larvae, often climbing into trees to reach the hives. An association has evolved between the honey badger and a small bird, the honeyguide, in which the bird flies just ahead of the badger, leading it to wild bees' nests. Using its strong claws, the badger then rips open the hives and both partners share the spoils.

It will also use its powerful claws to rip open termite mounds to get to the insects inside.

A thick layer of subcutaneous fat protects the honey badger from bee stings. Another remarkable characteristic is that its skin is so loose on its body that the badger can twist around in its skin and bite an attacker which might have a hold on the back of its neck.

■ If surprised by a powerful predator, with no escape possible, a honey badger, like this one in South Africa, will feign death.

WILD BOAR
Sus scrofa

THIS ancestor of the domestic pig is usually found in forest and thick undergrowth. Its thick skin allows it to penetrate dense thickets that no other animal would enter. Its winter coat is especially coarse and bristly. Active mainly in the night and early morning, the boar rests during mid-day. It lives in small family groups of up to twenty. However, old boars remain solitary.

The male has prominent tusks derived from canine teeth. Omnivorous, it forages over fairly wide areas when digging for bulbs and tubers, eating nuts and other plant materials, and also small animals and carrion.

It wallows regularly to remove parasites and also to form a mud coating which, in hot climes, acts as a sunscreen. The wild boar is an agile, fast-moving animal that becomes quite aggressive when threatened. Males will use their strong tusks in defending themselves. When endangered, the adult female signals the alarm with a "wuff" cry and the young instinctively lie motionless in cover.

Each wild boar piglet has its own specific teat from which to suckle and those born first choose a teat near the sow's head where, attracting her attention, they have less chance of being trodden upon.

As the group moves along in the Assamese forest, a female wild boar will grunt rhythmically to signal the young to stay together.

CAPYBARA
*Hydrochaeris
hydrochaeris*

. . . .

THE largest living rodent, growing to the size of a small pig, the capybara has a particularly large head, forefeet with four toes and hindfeet with three toes, all of which are partially webbed.

Primarily diurnal, it spends much of its time eating aquatic vegetation, an activity which is usually done in small groups.

A capybara is an excellent swimmer, moving through rivers and lakes with only its eyes, ears and nostrils showing above the surface of the water.

Its cheek teeth continue to grow throughout its life to counteract the constant wearing away caused by chewing.

The capybara has been extensively hunted for its hide and its meat and that threat, as well as the reduction of its natural habitat as it is drained for cattle pasture, has made the capybara so shy that it will flee at any strange noise or movement.

■ The capybara, like this one in Brazil's Pantanal, is a quiet, intelligent animal that rarely fights, either other capybaras or even possible predators.

GERENUK
Litocranius walleri
....

HIS large gazelle with its exceptionally long neck (thus its Somali name which means "giraffe-necked") and very elongated limbs is able to reach high up into tall bushes by standing erect on its hind legs. The male is distinguished from the female by its horns (the female has none) and the female's dark patch on the crown.

Living singly or in small groups the gerenuk browses on prickly bushes and small trees, using a foreleg to pull down the branches, eating the more tender leaves. A gerenuk requires no water, getting sufficient from the foliage on which it feeds. When running, it holds its head forward and at the same level as its body.

Its principal predators are cheetahs, leopards, lions, hyenas and wild dogs.

■ Gerenuks, such as these in Samburu, eat only the foliage of bushes and trees, never grasses.

RHESUS MACAQUE
Macaca mulatta

An Old World monkey, the rhesus macaque belongs to that family of monkeys that eats fruits, seeds, berries, insects and other small creatures rather than feeding primarily on leaves. It is one of the more plentiful macaques, although its range in many rural areas is declining rapidly.

It travels about during daylight hours either alone or in small groups of family units, each such unit headed by the founding female. It is a territorial animal and will defend its territory from intruders; however it is only very rarely that subordinate groups will challenge dominant groups to physical battle.

The rhesus macaque is equally agile on the ground and in trees where it sleeps. It is an intelligent animal and has a strong sense of curiosity. Learning quickly to manipulate simple tools, this is the species that was the organ-grinder's monkey.

■ Even though the rhesus macaque, this one is in Borneo, is comfortable on the ground. it rarely ventures more than a few yards from trees.

HARP SEAL

Pagophilus groenlandicus

THE harp seal is so named because of the dark U-shaped marking on its back. The contrasting colors are far more apparent in the adult male than the female. However, newborn pups of either sex have an overall white coat which, by the end of the second week, begins to turn mottled gray.

At some time between the middle and end of February groups of pregnant females congregate on the ice, each keeping its own distance from the others, to give birth to their young.

Harp seals may have the shortest nursing period of all mammals, about ten days, after which they are abandoned by their mothers who go off to mate. The males are noisily awaiting the arrival of the females and will drive off rivals by biting and hitting with their flippers.

A rapid swimmer, the harp seal can also move quickly over icy surfaces. However, one of its most useful traits is its ability to stay submerged for up to fifteen minutes when it plunges deeply down into the water for food.

■ In the raucous din of the squealing pups and barking adult females on Canada's Iles de la Madeleine, each cow somehow identifies her pup's cry from all the others.

BRAZILIAN TAPIR
Tapirus terrestris

THE tapir has a long proboscis formed by the upper lip and nose. This gentle, timid animal lives in forests, almost always close to water to which it takes flight at the first sign of danger. In addition to its mobile snout, it also has flexible ears, somewhat like those of a horse. It has four toes on each forefoot and three toes on each hindfoot. The Brazilian tapir is the smallest of the species.

While the adult's coat is an overall dark brown, without folds, the newborn are striped and spotted.

A vegetarian, it exists mainly on wild fruit, leaves, buds, shoots, and aquatic plants. Additionally, it will occasionally invade cultivated fields. It is a good swimmer and diver, attributes which it uses in fleeing from jaguars, its principal enemy. It is also a fast runner on land, even over rugged, mountainous country.

It frequently coats itself with a crust of mud as protection against insect bites. Normally silent, it grunts when disturbed.

■ Although protected, the Brazilian tapir is nevertheless still hunted for its meat.

KILLER WHALE
OR
ORCA
Orcinus orca

LARGEST of all the dolphins, the killer whale is easily identified by its conspicuous large dorsal fin and the black and white patch behind its eye. It is an avid predator, the only whale that preys on other warm-blooded animals including other whales and dolphins, yet it has never been known to attack a human being in the water. Orcas travel in extended family pods which cooperate in hunting.

A killer whale has forty to fifty sharp, interlocking teeth which point to the back of its throat. This enables the whale to tear big chunks from large prey. Its powerful jaws are so large that it can swallow a small seal whole. It can toss prey weighing up to 650 pounds into the air.

It has no fear of any predators and appears to be unafraid of approaching ships. An orca can swim at speeds up to 35 miles an hour and can make spectacular leaps out of the water. The splash resulting from these leaps can be heard for up to two miles.

Although it is capable of deep dives, the orca usually remains close to the surface and is often seen lying motionless with its head just above the surface while observing its surroundings.

An orca male is much larger than the female. This one is off the coast of Vancouver.

CAIMAN
*Caiman
crocodilus*

$\mathcal{L}$IVING in slow, still waters, this member of the crocodile family has become an endangered species as the population of wild caimans continues to diminish drastically. It is hunted not only for its skin, but the young are captured and sold as pets or stuffed and sold as curios.

On land the caiman is quite ungainly and flees quickly when frightened. Its greatest agility is in the water, where it also finds most of its food: fish, other reptiles, capybaras and the occasional bird.

The caiman is well protected from predators by its thick, leathery skin. Its very sharp teeth, used for tearing apart prey, are gradually replaced as they wear out.

The female caiman normally lays between twenty and fifty eggs at a time, after first making a nest from plant debris.

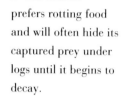

■ The caiman prefers rotting food and will often hide its captured prey under logs until it begins to decay.

October, 1991 - Pantanal, Brazil

Cutting through this marshy terrain is a dusty, rutted dirt track crossing over many boggy swamps. It is the path that leads to the fierce Pantanal caimans and I had come here to photograph them.

Just after crossing over a sluggish river and descending back to road level, I was amazed to see a huge gathering of caimans either lazing on the banks or swimming in the muddy water.

I set up my camera and tripod on a knoll partway down a hill and began photographing. A few, disturbed by my presence, wiggled into the water and swam away. A few glared balefully in my direction. Most remained unmoving.

As I stood there, dozens more arrived, the newcomers seemingly indifferent to my closeness.

Then one particularly large and brazen caiman started to slowly creep up the hill towards my position. That's when I picked up my gear and left.

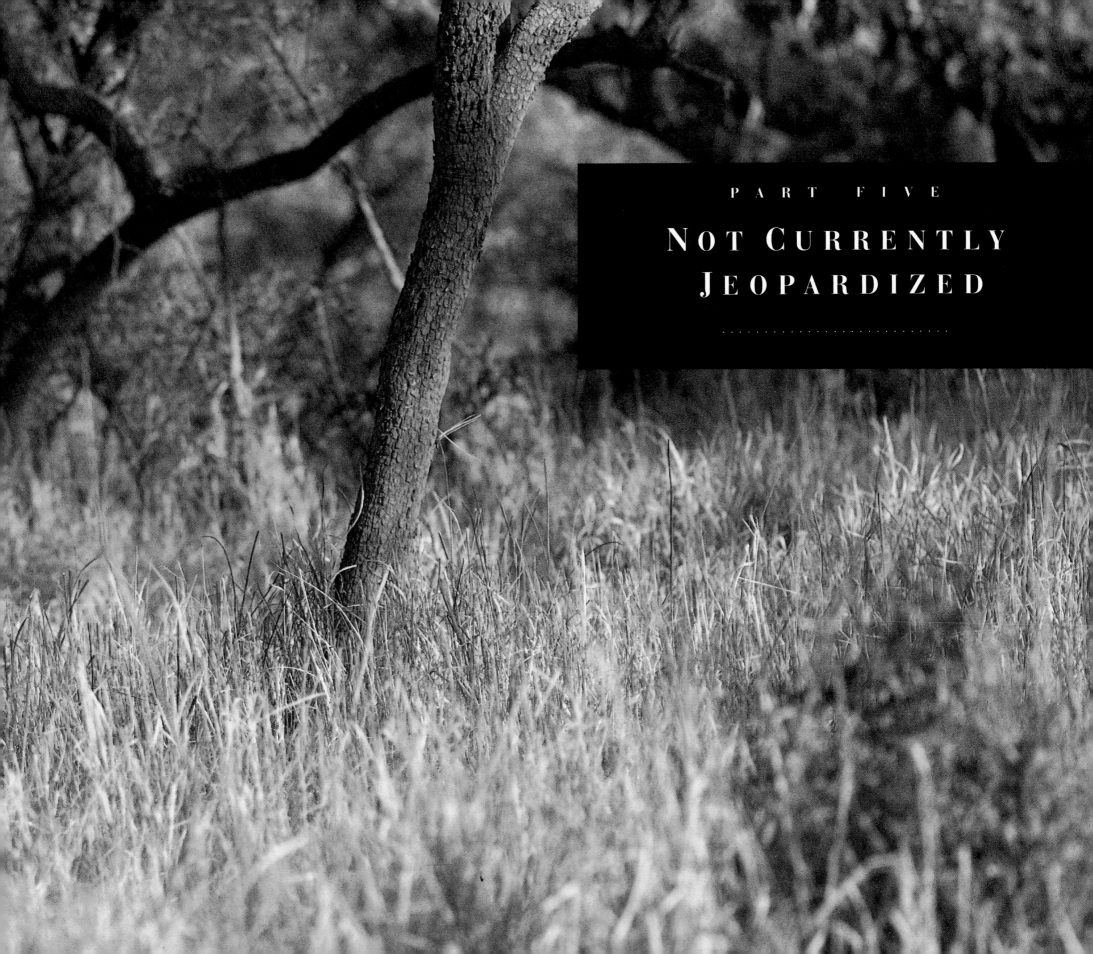

PART FIVE

NOT CURRENTLY
JEOPARDIZED

ANACONDA
Eunectes noctaeus

FOUND in marshy areas and sluggish, muddy waters, this enormous, voluminous snake grows to a length of 21 feet (which is somewhat smaller than the Amazon anaconda which can be as long as 30 feet). Although omnicarniverous, the anaconda prefers mammals as food.

It does not pursue its prey but waits for animals or birds to come to the edge of the water to drink. It then seizes its victim and kills it by coiling and constricting its own body around the captured animal. When necessary to fight a powerful prey, the anaconda finds a tree trunk around which it coils its tail in order to get a better hold before coiling its own body around the victim to be strangled. After the prey has lost consciousness the anaconda will slowly devour it, enlarging its own body to accept the larger sized victim. Later it may lie motionless for up to several days.

It is usually a night wanderer, moving about much less during daylight hours. It glides along with only the top of its head showing and can remain fully submerged for up to ten minutes.

In breeding season the males court their mates by issuing loud, booming sounds. The females produce litters of 15 to 30 young.

■ The anaconda is a boa, among the most primitive of all snakes, that kills its victims by crushing them in its coils. This one is in the Pantanal.

COATI
Nasua nasua

A member of the raccoon family, the coati is easily identified by its short legs, long, banded tail and long, tapered nose. Its coat, which is short on the legs and face, is longer on the body itself.

An omnivore, it lives in groups of up to forty individuals that hunt as a pack both day and night, resting in the heat of the day. It spends its nights in trees but may descend when frightened.

Holding its tail high and erect, the coati uses its amazingly mobile snout to probe in holes and cracks while foraging for the insects and spiders that are its main staple as well as several varieties of fruit of which it is particularly fond.

Pregnant females leave the group, going off alone to give birth, rejoining the group when the young are about two months old. Young coatis engage in constant, noisy play, chasing one another up and down trees. Older males leave the group and remain solitary. The social unit is structured on the single female with her young, which remain with her until they are about two years old. These females, with their young, often form large, wandering groups.

■ Aided by its prehensile tail, the coati, such as this one at Iguassu, is able to run up and down trees at great speed.

AXIS DEER
OR
SPOTTED DEER
Cervus axis

THE white markings of the spotted deer are retained throughout its entire life. The three-tined antlers, present only in the males, are shed annually.

Living in woodlands and forests, the herds consist of two or three males and many females. Grazing in the early morning and late afternoon, it finds cool places to rest during the heat of the day.

Its main predators are tigers and leopards. When threatened by these or other potential enemies the spotted deer takes refuge in streams or rivers as it is an excellent, fast swimmer and can usually escape capture by swimming rapidly to safety.

■ In India, spotted deer often wander through human settlements without fear.

IGUANA
Iguana iguana

. . . .

HE iguana is a New World diurnal arboreal lizard found in tropical forests. The full length of its spine is crested with a row of comb-like spines, higher at the neck and receding to its lowest at the tail. Its tail can be up to three times the length of its body. Although tree-dwelling and a superb climber, this agile lizard is also an excellent swimmer.

The colored bands that circle the body and tail of a young iguana may be very light in color but darken as it gets older and the contrasting colors then become quite distinct. It feeds mainly on leaves, berries, fruit and other plant food but it will also eat small mammals and nesting birds.

Although its diet is primarily vegetarian, the iguana has very sharp teeth and claws which are used most effectively in defending itself against predators such as large cats, caimans and boas. Additionally, its long, curved claws give it an excellent grip on branches when climbing in trees.

■ Dragon-like in
appearance this
iguana, seen on the
road in Argentina, is
actually a timid
reptile.

GREATER KUDU
Tragelaphus strepsiceros

THE greater kudu bull is likely to be a solitary animal as the usual herd consists of small family groups of females and young without adult males. Mainly nocturnal, a greater kudu rests for most of the day in shady areas, preferably on high ground.

It feeds on a wide variety of leaves and shoots, including poisonous plants which it can apparently eat without harm. It is a superb jumper and can easily vault over objects six feet high. It has a heavy, clumsy running gait which throws the tail up into the air, exposing its white underside.

Although very sensitive to sound, the call of the greater kudu is the loudest of all antelopes and, during the mating season, a bull's roar is heard over a great distance.

The greater kudu has majestic curving horns which, in older males, spiral round two and a half times. The horns are used in ritual fighting among the males, the two opponents locking horns and wrestling until the weaker animal concedes defeat. It is the beauty of the greater kudu's horns, much prized by trophy hunters, that has led to a great decline in the number of these creatures.

■ During the breeding season an adult male greater kudu like this one in the Transvaal will fiercely defend his territory against rivals.

LESSER KUDU
Tragelaphus imberbis

THE lesser kudu is smaller and more graceful than the greater kudu, having smaller horns which form a closer spiral and are less diverging and stripes which are more numerous and contrasting. The lesser kudu frequently inhabits much drier land than the greater kudu.

It has good scent and hearing but poor eyesight. Eating mainly the vegetation of trees and bushes, it also eats grass, fruits and roots in smaller quantities. It is able to go for a month without water.

The lesser kudu lives in small groups of not more than five or six animals, usually in the vicinity of an old male. A very timid creature, it spends most of its daylight hours hidden in dense brush moving out most warily only after dark. Its principal predators are leopards, lions and wild dogs.

■ Throughout southern Africa, young lesser kudu calfs are preyed upon by baboons and pythons.

NILGAI OR BLUE BULL
Boselaphus tragocamelus

THE nilgai lives in open plains and has some characteristics of ox, deer, goats and camels. It is the largest of the Asiatic antelopes. Its forelegs are slightly smaller than its hindlegs and it has a long, pointed head. Only the male has horns.

Females and calves live in herds but male nilgai are usually solitary or in small parties. They are browsers but also eat fruit and have often done considerable damage to sugarcane plantations.

Within its own marked territory, the nilgai establishes paths which are used to reach watering holes, resting places, etc. During mating season, which occurs in March, when fighting each other for available females, nilgai bulls get down on their knees to do battle.

■ The nilgai, called a "blue cow" or "blue bull" is a slate blue antelope found only in India.

NYALA
*Tragelaphus
angasii*
. . . .

THIS slenderly built, narrow-bodied antelope travels in small groups, usually females and young with a single bull. Exclusively a browser, it feeds through the night, from late afternoon until early morning, but it moves about as well during most daylight hours.

Its clumsy gait is similar to the greater kudu and when it runs its tail is thrown up into the air, exposing its white underside.

Fond of fruit, it will frequently browse under trees containing baboons and monkeys so that it might eat the fruit that has been dislodged and fallen to the ground. However, as baboons are among those animals that prey on small nyala, young nyala are at risk.

■ The nyalas, such as this one found in the Transvaal, have good scent and hearing, but poor sight.

SAMBAR
Rusa unicolor

. . . .

THE long-coated bristly-haired Indian sambar is a solitary deer in which only the stags have antlers, usually with three tines on each.

It eats mainly grass, leaves and wild fruit and is often in marshy areas feeding on aquatic plants. As a rule, it is not easy to catch a glimpse of a sambar because, even in regions where it is quite numerous, it will dart off into dense forest at the slightest sound.

The sambar is a good swimmer and, if chased or attacked by a tiger, leopard or wild dogs, it will jump into the nearest river or stream and swim quickly away.

The stags acquire large harems during mating season and will then defend them most vigorously from possible rivals.

■ Sambars are large Asiatic deer that usually live in open, deciduous forests.

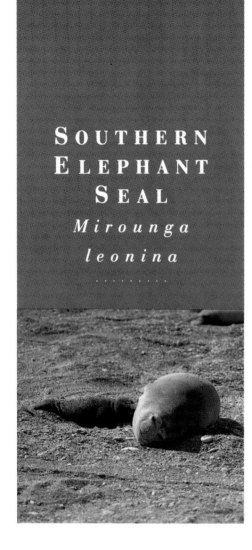

SOUTHERN ELEPHANT SEAL
Mirounga leonina

THE southern elephant seal found in South America's coastal waters is the largest of the pinnipeds (marine animals having flippers: seals, sea lions, walruses) and may weigh up to 8000 pounds. It is a huge blubbery beast that lives out most of its life in the sea. During breeding season, three-ton bulls, mountains of fat, slam and slash one another in the battles for females. The females are only about one-third the weight of the males and they do not have the pendulous snout that gives this breed its name. The elephant seal has no external ear structure.

A single bull may have a harem numbering as many as 100 cows which he defends continuously as a bull must constantly prove himself in battle to be superior to competing bulls. As a result of these battles, many of the bulls are heavily scarred, particularly around the neck area.

New born pups weigh about 80 pounds and then gain about 20 pounds per day during the nursing period which lasts just over three weeks.

■ A male southern elephant seal such as this one in Patagonia is old enough by six years of age to command his own harem.

TAMANDUA
Tamandua tetradactyla

AN arboreal anteater, the tamandua has a prehensile tail which it uses to assist it in climbing and to strengthen its grip. Its fur, a mottled pattern of off-white and dark yellow, provides excellent camouflage for the tamandua as it hides among the leaves while hunting for food.

From the end of its very long snout, the anteater extends its rounded sticky tongue up to sixteen inches to feed on termites, tree-living ants, beetle larvae and bees, all of which it detects by scent. In order to get to its food, the anteater uses its strong claws to rip open the insects' nests. In defending itself it will use those sharp claws and its powerful arms to strike and tear at an enemy. Usually, if it senses danger, the tamandua will try to escape. If this is not possible, the tamandua will sit down and try to hug its adversary.

In order to prevent its long, pointed claws form digging into its palms, the anteater walks on the outside of its hands. It is quite clumsy on the ground and spends most of its life in trees.

The very young anteater is carried on its mother's back until it is able to fend for itself.

■ Unlike the giant
anteater, the
tamandua or lesser
anteater, like this one
discovered in Belize,
spends its life in the
trees.

TAYRA
Eira barbara

A slender carnivore that lives alone, temporarily pairing only during mating season, the tayra is active both day and night, rarely resting. It feeds on guinea pigs, mice and squirrels. It also preys on domestic poultry, but only when other food is not available as it prefers to stick to the wild.

With its large, blunt claws, it is an excellent climber, both up and down trees. Webbing between its toes gives it extra impetus when swimming. On land, it hides in hollows or among dense thickets of branches as well as in the burrows of other animals, particular rodents, which it first kills. Local people sometimes keep tayras to help control rodents in their homes and gardens.

Extremely territorial, tayras have even been known to hiss and snarl at human intruders, although outright attacks are unlikely.

The tayra is a particularly bold and voracious predator in South and Central America.

WATUSSI
Bos primigenius

○RIGINALLY from the plateau region of Uganda, this impressive breed of large cattle is remarkable for its immense horns.

In rare instances a calf is born with a white forehead. In earlier times, all such specimens belonged automatically to the chief of the Ankole tribe. It is probably for that reason that this animal is also known, in some regions, as an 'ankole'.

This diurnal vegetarian is used as a beast of burden in its native land and is also raised for its milk which is a staple food among the tribespeople of the area and its hide which produces excellent leather.

■ Man's first ally in agriculture was the draught animal; in parts of Africa cattle such as the watussi still provide the muscle.

*T*HE last entry in this volume is, perhaps, the strangest of all, in addition to differing in so many ways from every other creature photographed.

All the species in this book have been reasonably large and even the smaller creatures are of sufficient size to be easily spotted at some distance. This one is so tiny that it is almost invisible until only a few inches separates the viewer from the subject.

Although the book contains examples of many different mammals, reptiles and birds, all have one common feature - they are vertebrates. Every one of them has a spinal cord. Unlike all these others, this last depicted animal is an invertebrate.

Every other animal in this book is hidden from human view by the dark of night. This one is most visible after dark. How incredible is the limitless diversity of nature!

September, 1991 – Das Emas, Brazil

I had heard tales about luminous termite mounds that glowed eerily in the dark. It was supposedly a phenomenon that happened only rarely, only in a few of the billions of termite mounds in the region, under very specific climatic conditions.

I knew no one who had actually seen one glowing nor was there anything in the literature. I considered it a "mirage."

That is, until I saw it and photographed it.

Just before midnight, looking out through the van's window, I saw a faint glow. Using a flashlight, picking my way among the bushes, I came suddenly upon one mound from which many tiny lights glowed. I could hardly believe my eyes and quickly photographed the astonishing display.

I then returned to the van, used its headlights to illuminate the mound, and re-photographed it from the same position.

I have subsequently learned that the lights in the termite mounds are caused by fluorescing click beetle larvae.

CLICK BEETLES
(bioluminescent in termite mounds)

Elateridae pyrophorus

. . . .

THESE New World luminescent click beetles, still in the larvae stage, enter into selected termite mounds to prey upon junior and adult termites. The beetles use their phosphorescent glow, emitted from luminous organs on the underside of their abdomens, to attract the termites so that they can devour them. They time this to coincide with the swarming of adult winged termites, an event which seems to occur only after a long period of hot, dry days and nights. In the dark, the brightly luminescent larvae, sometimes referred to as "glow worms", can then be seen through tiny openings in the sides of the termite hills. These are the exit holes for the winged termites that must fly out to mate and go to the ground to lay their eggs. Interestingly, the beetle's light mechanism is remarkably efficient, for about ninety percent of the energy radiated by its photogenic cells is in the form of visible light whereas, for example, an ordinary light bulb has only about five percent. Unlike the intermittent flickering light of a lightning bug, the click beetle's phosphorescence is a comparatively steady glow.

*A*ND so, at the end of my book, I pose three questions:

● How astonishing are the wondrous marvels of nature?

● How much can they add to the richness of our lives?

● How effective can we be in preserving our planet and its biodiversity?

Only you, the reader, can provide some of the answers.

ANDERSON, S. *Simon & Schuster's Guide to Mammals.* New York: Simon & Schuster. 1982.

BURGER, C. *Mammals of Southern Africa.* South Africa: 1987.

BURTON, J .A. and PEARSON, B. *Collins Guide to Rare Animals of the World.* London: Collins. 1987.

BURTON, M. *The New Larousse Encyclopedia of Animal Life.* New York: Bonanza Books. 1967.

CORBET, G. B. and HILL, J. E. *A World List of Mammalian Species.* London: British Museum (Natural History). 1980.

COX, T. *Travellers Guide to East Africa.* London: Thornton Cox.1980.

DARLING, J. *Wild Whales.* Vancouver, Canada: Summer Wild 1987.

DORST, J. and DANDELOT, P. *A Field Guide to the Larger Mammals of Africa.* London: Collins. 1976.

EVANS, P. R. *The Sea World Book of Seals and Sea Lions.* NewYork: Harcourt Brace Jovanovich. 1986

FETNER, P. J. *The African Safari.* New York: St. Martin's Press. 1987.

FITTER, R. S. R. *Vanishing Wild Animals of the World.* London: 1968.

GROBLER, H., HALL-MARTIN, A. and WALKER, C. *Predators of Southern Africa.* Johannesburg: Macmillan 1984.

HALTENORTH, T. and DILLER, H. *A Field Guide to the Mammals of Africa.* London: Collins. 1977.

HOYT, E. *The Whale Watcher's Handbook.* Ontario, Canada: Penguin. 1984.

MCBRIDE, C. *The White Lions of Timbavati.* London: Paddington Press. 1977.

MENDELL, E. *Wildlife Odyssey.* Windlesham, U.K.: 1990.

MERZ, A. *Rhino at the Brink of Extinction.* London: Harper Collins 1991.

PAYNE, J. and ANDAU, M. *Orang-utan, Malaysia's Mascot.* Kuala Lumpur: Berita. 1989.

PERRINS, C. M. and MIDDLETON, A. L. A. *The Encyclopaedia of Birds.* London: George Allen & Unwin. 1985.

PRESTON-MAFHAM, R. and K. *Primates of the World.* London: Blandford 1992.

RAVAZZANI, C. FILO, H. W., FAGNANI, J. P. and DA COSTA, S. *Pantanal, Brazilian Wildlife.* Curitiba, Brazil: 1990.

SINCLAIR, T. *Introduction to India.* Hong Kong: Odyssey Guides. 1991.

SMART, T. *Tropical Rainforests of the World.* Godalming, U.K.: The Book People. 1990.

SMITHERS, R. H. N. *The Mammals of the Southern Africa Subregion.* Pretoria: University of Pretoria. 1983.

STEVENS, K. *Jungle Walk.* Belize: Angelus. 1987.

STUART, C. and T. *Field Guide to the Mammals of Southern Africa.* Cape Town: Struik Publishers. 1988.

STUART, C. and T. *Southern, Central and East African Mammals.* Cape Town: Struik Publishers. 1992.

STUART, C. and T. *Guide to Southern African Game & Nature Reserves.* London: New Holland Publishers. 1992.

U.S. DEPARTMENT OF THE INTERIOR, U.S. Fish & Wildlife Service. *Endangered & Threatened Wildlife and Plants.* Washington: 1989.

WALKER, E. P. *Mammals of the World.* Baltimore: Johns Hopkins Press. 1975.

WHITFIELD, P. W. *Macmillan Illustrated Animal Encyclopedia.* New York: Macmillan. 1984.

WORLD CONSERVATION MONITORING CENTRE *1990 IUCN Red List of Threatened Animals.* Cambridge, U.K.: IUCN Publications Services Unit. 1990.

YOUNG, J. Z. *The Life of Mammals.* London: Oxford University Press. 1957.

INDEX